P9-CAH-982

THE EARTH BOOK

THE EARTH BOOK

by Gary Jennings

J. B. LIPPINCOTT COMPANY
Philadelphia and New York

Photograph on p. 25 (bottom) by Reagan Bradshaw. Photograph on p. 6 courtesy National Aeronautics and Space Administration. Photograph on p. 26 by Jim Olive. Photographs on pp. 2, 5, 8, 9, 10, 11, 12, 15, 18, 19, 20, 21, 22, 23, 24, 25 (top), 27, 28, 29, and 30 by Blair Pittman. Photographs on pp. 7, 13, 14, and 17 by the author.

U.S. Library of Congress Cataloging in Publication Data

Jennings, Gary.
The earth book.

SUMMARY: Examines the many different kinds of pollution to which man is subjecting the earth and outlines some steps the reader can take to deal with this problem.

1. Ecology—Juvenile literature. 2. Pollution—Juvenile literature. [1. Ecology. 2. Pollution]
I. Title.
QH541.14.J45 301.31 74-8405
ISBN-0-397-31268-7

Printed in the United States of America
First Edition

You are one of the first human beings ever to see the planet Earth entire. Of the hundred billion people who lived before your time, none ever saw more than a tiny fragment of the world at once. But now, through the camera eyes of satellites and spaceships, you can view our world "in the round"—the bright blue green of its waters, the rich browns of its land areas, all marbled with the white swirls of clouds. It looks both lovely and lonely. And lonely it is, for we know of no other planet like it. We know of no other planet that has air to breathe, water to drink, and topsoil to support green growing things. Earth has all three.

Without any one of them—air, water, topsoil—all life would disappear: insects, fish, flowers and trees, animals, birds, and man. Our planet would become as dreary, barren, and lifeless as that cinder we call the moon.

The earth's atmosphere, or surrounding envelope of air, is the very breath of life for everything that lives, from plants and birds to you and me. To us here on the ground, the atmosphere looks to be "as high as the sky." But from space one can see that the atmosphere enveloping earth is as thin as an apple skin is to the apple.

Sad to say, ever since man began to develop a civilization, he has been thoughtless in his treatment of that thin and delicate envelope of air. He has been polluting it (fouling and poisoning it) for thousands of years. This pollution began mildly enough with the smoke from the cavemen's cooking fires. But now, with the exhaust from motor vehicles and the smoke and fumes from factories, air pollution has become a drastic problem.

Like the earth's air, its water supply is vital to all forms of life. Scientists still do not know exactly how it came about that our world's atmospheric gases combined in such a fortunate way as to produce water, but they do know that earth is the only planet orbiting our sun with enough water to support life. Until recent times, most of our earth's water was as clear, clean, and sparkling as you see it here.

But, over the ages, man has disposed of almost everything he wanted to get rid of by dumping it into the nearest body of water. Into that once-clean water he has flushed his toilets, thrown his garbage, poured the wastes from his factories. And today much of the world's water no longer dances and sparkles but crawls with sludge and slime.

The earth's topsoil is a mixture of dirt and nutritious chemicals that feeds and supports every green growing thing, from these tiny Indian pipe plants to the most gigantic trees, including all the vegetable foodstuffs of animals and men. The earth's blanket of topsoil took millions of years to develop, but it is far thinner even than the planet's envelope of air—and is even more easily harmed by careless treatment.

Over the centuries, man has gradually cleared the earth's land areas of much of their plant cover. Without its "umbrella" of plant cover, the precious topsoil is laid open to erosion. Rain falling on it cuts first gullies, then ditches, washing the rich topsoil away to the bottom of a river or lake. There it is useless for growing anything. What is left behind, after erosion, is only the unproductive subsoil, as barren and lifeless as desert sand.

Of course, the earth endured some natural erosion long before man had any hand in causing it. A million years ago, the swift-running Colorado River began to carve the mile-deep Grand Canyon of Arizona, shown here. And the earth has always endured other natural influences. The air has forever been mildly "polluted" by the pollen scattered by flowers and the salt thrown up by ocean waves. Rivers have forever been mildly "polluted" by nature's garbage, fallen leaves and the like. But with the coming of man, the face and features of the earth *really* began to change, and not always for the better.

The first men on earth were roving hunters, and they left no more mark on the land than did the beasts they hunted. But, as the ages passed, some men began to settle down, to become farmers and shepherds. They cleared the trees and brush from pieces of land to make room for their houses, herds, and crops.

While some men continued to live well apart from each other on their farms or ranches, others began to bunch together for the convenience of trading with one another or for mutual safety from unfriendly tribes. Their close-together houses became villages and small towns—and the earth is still well speckled with these small communities.

In time, numbers of the small towns prospered, attracted more settlers, and grew to be medium-sized cities. There are still many of these, too, on our earth.

And often a small town grew eventually into the monster-sized city that we call a metropolis—sprawling in area, towering with skyscrapers, and teeming with millions of people.

However, during those same ages when more and more people were bunching together, still other men drifted *away* from the centers of civilization. These explorers and pioneers dared to penetrate unknown and unopened lands, and they were followed by others who came to settle.

Now there is scarcely a spot on earth that has not known the footstep of man. There are towns and cities where once there was only wilderness.

But it was not until the past two centuries, when man began to depend so heavily on technology—science and machines—that his marks on the earth became scars and his influence on nature became harmful. For example, the worst features of civilization—dirt, smell, litter, noise, overcrowding—used to be confined to the cities.

The invention of the automobile made it possible for almost anybody to travel almost anywhere, and to take pollution with him. Long strips of land are now buried under concrete or asphalt for cars to run on. Once-quiet places now echo to the noise of engines and the screech of brakes. Once-pure air is fouled by the cars' exhaust gases. Once-private places are no longer, and all the wild creatures that lived there have perished or fled.

Because of the automobile, man's cities spread outward, like a stain, into once-open lands. Advertisers are eager to appeal to the millions of drivers always on the move, so they clutter the roadsides with "eye pollution" like this. The billboards and neon signs may extend from one city all along a country highway clear to the next city's similar cluster of clutter.

MOTOR HOTEL
NITE CLUB
RAMADA INN
PIT BAR-B-Q
STATE
DINING ROOM
FOR SALE
5.3 ACRES
CALL OWNER 526-1711
CHEVROLET

Lately, another kind of machine has become even more of a plague than the automobile. This is the off-road vehicle—the trail bike, the dune buggy, the snowmobile, and the all-terrain vehicle (ATV). Trail bikes spin their wheels as they climb slopes, cutting gullies that start erosion of the topsoil. Snowmobiles mash seedlings that would have become trees.

All of the off-road vehicles panic birds and animals, crush nests and eggs, and bring noise, smoke, smell, and filth into the last few wilderness places where wild creatures (and people tired of machines) might have sought seclusion.

As man's technology has made him increasingly dependent on machines, especially those that provide transportation—cars, airplanes, ships, snowmobiles—the demand for metals to construct them and fuel to run them has increased. When our needs were more modest, a mine was no more unsightly than a hole in the ground. But now the "more productive" methods of strip mining and open-pit mining tear up all the surface terrain around a deposit of coal, copper, or other mineral. Note the size of this open-pit mine's giant power shovel, as compared with the truck and bulldozer alongside. As the monster shovel scrapes and digs, it leaves behind it a ravaged, gray brown, ugly "moonscape" that may extend for many square miles.

While the world's once-open lands continue to shrink before the spread of man's civilization, they are also cluttered with the waste products of that civilization. Every year, millions of tons of garbage are piled in city dumps like this one. And other millions of tons of trash are simply scattered as litter across the countryside or carelessly tossed into the nearest body of water.

The automobile also makes its contribution to "eye pollution" when it is worn out and must be disposed of. On the outskirts of almost every town, we see ugly "automobile graveyards."

Our homes and factories would not function without electricity, so thousands of hydroelectric dams have been built to convert water power into the needed electric power. But, in too many cases, the waters that back up and become lakes behind the dams have submerged the living places of wild animals and birds. And many kinds of wildlife, like the alligator, are in danger of disappearing altogether because their homes and food supply have been disturbed.

Other wild creatures get killed more or less by mistake. Man protects his farm crops from insect pests by spraying insect-killing chemicals like DDT. But then birds eat the poisoned insects and die themselves—like the cardinal shown here. DDT is now banned from use in the United States, but already America's bird population has been sadly diminished.

Our major rivers have for centuries been shipping lanes, and modern engine-powered riverboats foul the waters with oil, grease, and wastes dumped overside.

Similarly, sports motorboats on our lakes and rivers leave oil slicks on top of the water and pump their exhaust gases under and into the water. All these pollutants eventually make the waters unfit to drink or swim in and kill their fish and other aquatic life.

Even the mighty oceans can swallow only so many insults in the form of pollution. They must absorb all the filth brought by the rivers that flow into them. And nowadays they must also endure vast oil slicks spilled by wrecked tanker ships or by offshore oil wells that run wild (like this one, which caught fire and belched millions of gallons of oil before it could be capped). Scientists are worried that ocean pollution may kill off many of the world's most important food fish and even the ocean plants called plankton, which manufacture much of our atmosphere's life-sustaining oxygen.

Many of the earth's cities, on many days every year, don't have pure air to breathe—they have *smog*, a word made up of "smoke" and "fog." Smog contains gases that irritate one's nose, throat, and eyes; acid chemicals that eat at skin, clothing, and other materials; and even particles of poison that can kill small children or ailing and elderly adults. Smog is most likely to occur in industrial communities where factory chimneys pour their fumes into the air. But almost every sizable city today, because of the fumes from its heavy automobile traffic, has the same problem.

Like automobiles, jet airplanes also leave trails of smoke and fumes. But the chief pollution hazard of jet planes may be their contrails of condensed water vapor, which we see as long, narrow white clouds against the sky. The contrails *are* clouds—artificial clouds. Scientists fear that, as more and more planes leave more and more of these clouds in the upper atmosphere, much of the sun's life-giving energy may be cut off from the earth.

The pollution and uglification of our once-beautiful earth can be blamed on many things—man's selfishness, greed, laziness, and ignorance. But these factors boil down to just one basic cause: too many people, and those people demanding more from the earth than it can provide. There is reason to fear that eventually the earth will not be able to support man in the manner he would like. There is a limit to how much food the earth can grow to feed our bodies and how much oil the earth can provide to fuel our machines.

But, long before we run out of food and fuel, we will have run out of other things that make life worth living—elbow room, privacy, independence, quiet, and peace of mind.

Now . . . what can *you* do about the pollution, uglification, and overcrowding of our earth?

The first thing is to recognize what's happening—and this book is only a first quick look around at some of the dangers facing our world today. Read all you can find on the subject. Look to your parents, teachers, and scoutmasters for advice on what you can do to help improve the planet.

For example:

- Don't spend your allowance to buy products made by any company which you *know* is helping to pollute the world. Try to stop your friends and family from buying those products. The one thing that can frighten any company and make it mend its ways is a threat to its money-making.
- Don't hesitate to write letters to such companies—or get someone to write them for you. After all, *you* are "a customer of the future," and you will be listened to.
- Write also to your mayor, governor, or congressman when you have a suggestion (or a criticism) about their handling of the world's problems. Again, *you* are "a voter of the future," and you will be listened to.
- Try to stop your family's buying any products that are *obviously* pollutants, such as aluminum cans and plastic containers (which never decay), throwaway bottles, and powerful insect killers and weed killers.
- When you have to go somewhere, get permission to walk or bicycle, rather than beg for a lift in the family car.

- When you can, help organized groups like the Scouts in their pickup and cleanup campaigns.
- Don't buy, and try to persuade others not to buy, any product that is advertised on billboards.
- Above all, don't be a polluter yourself!
  Save your litter for the litter basket.
  Put a litter bag in your family's car or boat.
  Pick up after your picnic.
  Walk your dog in places where people don't walk.
  Use an earphone when you listen to your transistor radio in public places.
  And do whatever you can to convince young people and uncaring grown-ups to do the same.

Perhaps, with your help, we have some hope of bringing back to life and loveliness the wonderful world that once we had.

GARY JENNINGS was born in the United States and is now living in Mexico. He is the author of several books, including *The Killer Storms* and *The Shrinking Outdoors*.